ROWDY S

Abhijit Naskar is a celebrated Neuroscientist, Bestselling Author of 100+ books, and the World's First Poet with 1000+ sonnets, who has been serving at the forefront of humankind's struggle against hate, intolerance, bigotry and fanaticism.

ROWDY SCIENTIST

HANDBOOK OF
HUMANITARIAN SCIENCE

ABHIJIT
NASKAR

Rowdy Scientist: Handbook of Humanitarian Science

Copyright © 2023 Abhijit Naskar

This is a work of non-fiction

An Amazon Publishing Company, 1st Edition, 2023

Printed in the United States of America

ISBN: 9798853321953

Also by Abhijit Naskar

The Art of Neuroscience in Everything
Your Own Neuron: A Tour of Your Psychic Brain
The God Parasite: Revelation of Neuroscience
The Spirituality Engine
Love Sutra: The Neuroscientific Manual of Love
Homo: A Brief History of Consciousness
Neurosutra: The Abhijit Naskar Collection
Autobiography of God: Biopsy of A Cognitive Reality
Biopsy of Religions: Neuroanalysis towards Universal
Tolerance
Prescription: Treating India's Soul
What is Mind?
In Search of Divinity: Journey to The Kingdom of Conscience
Love, God & Neurons: Memoir of a scientist who found
himself by getting lost
The Islamophobic Civilization: Voyage of Acceptance
Neurons of Jesus: Mind of A Teacher, Spouse & Thinker
Neurons, Oxygen & Nanak
The Education Decree
Principia Humanitas
The Krishna Cancer
Rowdy Buddha: The First Sapiens
We Are All Black: A Treatise on Racism
The Bengal Tigress: A Treatise on Gender Equality
Either Civilized or Phobic: A Treatise on Homosexuality
Wise Mating: A Treatise on Monogamy
Illusion of Religion: A Treatise on Religious
Fundamentalism
The Film Testament
Human Making is Our Mission: A Treatise on Parenting
I Am The Thread: My Mission
7 Billion Gods: Humans Above All
Lord is My Sheep: Gospel of Human
Morality Absolute
A Push in Perception
Let The Poor Be Your God
Conscience over Nonsense
Saint of The Sapiens
Time to Save Medicine
Fabric of Humanity
Build Bridges not Walls: In the name of Americana
The Constitution of The United Peoples of Earth

DEDICATION

To Nikola Tesla

CONTENTS

1. Rowdy Scientist Sonnet

Sonnet 1196

Rowdy Scientist Sonnet

Science is neither boastful nor bashful,
Science stands dutybound, forever mindful.
Science gives the final answer as a ray of hope,
With all avenues exhausted to the last granule.

Science is neither defensive nor offensive,
Science can't afford such primitive prerogative.
Transcending binary norms science acts whole,
Defying the comfort of all corrosive narrative.

Science is slave to none, science enslaves none,
Science blooms from reason, endorsing curiosity.
But don't ever confuse curiosity with cynicism,
Curiosity brings understanding, cynicism apathy.

Average scientists seek answers,
Great scientist seeks questions.
Submissive scientists chase solutions,
I'm a Rowdy Scientist, give me problems!

ABHIJIT NASKAR

2. I Make Gods

ABHIJIT NASKAR

Sonnet 1197

They make cars and rockets,
I make Gods and legends.
Cars and rockets take you places,
I shed light on mental mazes.

Science doesn't find the way,
Science makes the way.
Heart doesn't find the way,
Heart is the way.

Bugün hayat ne kadar zorsa,
yarın hatıralar o kadar tatlı olacak.
Hardship of today is tomorrow's sweetness,
Dolor de hoy es la dulzura de mañana.

Happiness doesn't mean checking all the boxes,
Happiness means to stop putting life in boxes.

3. Sin Amor No Hay Mente

Time is a slimy wimy thing,
Life is a blessy messy thing.
Love is a happy crappy thing,
Mind is a mighty fighty thing.

And yet,
Sin amor no hay mente,
Sin mente no hay vida.
Life is a record of love,
Time is a record of love -
To fathom this is to find felicidad.

Kendinize bir hediye ol,
Kendinize bir hayal bul.
Kendinize bir hasret ol,
Kendinize bir hayat kur.

No need to go look for a translator - I am yet to find a dictionary that can illuminate the depth, reaches and nuances of the turkish experience of "kara sevda" - or the vedantic expression "aham brahmasmi". The global realm of the human mind is far too vast to be confined into the narrow and ridiculously limited linguistic narrative of a single language.

That's why I say, if you wanna know about a culture, you can read about it in any language - but if you want to experience that culture like your own, you gotta do it as one of their own - through their own native language. Understanding the language is a quintessential part of understanding the culture.

4. Language is Highway

AI can translate info, not inkling,
AI can translate facts, not poetry.
Till a tongue transcends lips to soul,
Translations are but soulless forgery.

Each language leaves a distinct mental imprint,
Inaccessible by the fanciest of translator.
Translation gives a glimpse into the head,
Language is highway to the soul of a culture.

The sun doesn't know how to shine over only one planet, I don't know how to illuminate only one culture. What this means is that, it's not that I don't write from the narrow prehistoric confines of one single culture or tribe - I don't know how to. That's why I say, I am not a poet scientist who works on a paradigm, I am the paradigm itself. I don't have a message, I am the message.

5. I Am The Message

Sonnet 1198

I don't have a message,
I am the message.
I don't write to make sense,
I write to expand the senses.

I don't speak of the future,
Because I am the future.
I don't know how to write poetry,
that's why I am a poetic enigma.

I can't tell poetry and life apart,
Everywhere I see, my eyes find poetry.
I can't tell science and curiosity apart,
Wherever I look, a mystery beckons me.

I can't tell philosophy and wonder apart,
Good philosophy is just mental meanderings.
I can't tell divinity and kindness apart,
Practical holiness is a humanitarian inkling.

6. Duties to The World

Divinity organized is divinity lost - holiness organized is holiness lost - or simpler still - religion organized is religion lost. And it is the prehistoric priests of organized religion (not priests in general) who stand as mindless obstacle to science, sanity and universal sanctity.

They say, scientists are the new priests. Well, most priests leave things to god - we don't - we are gods. And not just us scientists - every human who takes responsibility for their society, is a living god. Even a priest can be god - those rare few who inspire their parishioners to be god-like rather than god-fearing. Whether you believe in the supernatural, that's irrelevant. The question is, are you mindful of your duties to the natural world?

The difference between creator and creation is in responsibility. The creator takes responsibility, the creation delegates it. What are you?

However, let me make one thing absolutely clear. Just like there are fear-mongering preachers in the world, there are also self-serving scientists in the world - both are the lowest form of life there is - because both abuse two of the most potent forces of nature, faith and reason, for personal gain.

That's why it's not enough to be a scientist, what the world needs are humanitarian scientists. The difference is - in practice, regular scientists are dedicated to science, and often they end up committing the worst kind of inhuman atrocities in the name of science, whereas humanitarian scientists are devoted to humanity, and science is merely their chosen means of service to humanity.

7. Science is Service

It's this simple - unless science and service become one, no science can do good. Science is service - if you don't get this, you are not a scientist, but a glorified con-artist - no matter how brilliant you are.

In the same way, till religion and oneness become one, no religion can do good.

Means may vary, but the objective must be one - human welfare.

Human intervention is divine intervention, and science is the most potent form of human intervention there is - and as such, it is the most impactful divine intervention.

Our problem is, we always try to draw a supernatural connotation in divinity. Thus all through history religion has provided a safe haven to all sorts of superstitions, some naive, others downright cataclysmic.

Throughout history religion has been peddled as the greatest excuse for indifference. Rather than empowering believers to rely on their own faculties, organized religion has made it a habit to peddle wishful inaction as the measure of religiousness. Because the more inactive they are, the more vulnerable they become, and the more vulnerable they become, the easier it is to manipulate them.

Now the point is, no matter how it has been till now - it can't go on like that. Believer or not, religious or not, humans gotta take charge of the human world, rather than cowardly waiting for some mythical messiah to drop from the sky.

Wanna change the world - first stop whining like backboneless fleshbags, and work till your last ounce of strength to attain a position of power. Run for office, be a copper, be a scientist, journalist, filmmaker - be something - anything - that earns you control over the paradigm. Infiltrate each and every corner of society that holds power, then use that power to lift up the world, towards an integrative, hateless and sustainable future. I'll be gone soon, but there'll come a time when every walk of society will be run by my soldiers. That day, I shall finally rest in peace, knowing that my world is in good hands - your hands.

8. Empty Yourself

In terms of tangibility
science is the greatest power.
In terms of sensibility
love is the greatest power.

Bravery without benevolence is imbecility,
Reason without affection is yet more rigidity.

Cansiz, Bir Şiir

Seni çok özlüyorum,
Ne zaman geleceksin ya!
Benim kafamı kırmaya,
Ne zaman bağıracaksın ya!

Ne olursa olsun hemen gel,
Sensiz hayat çok acıyor.
Gel, çıldırt bu bilim adamını,
Cansız akıllılık çok acıyor.

Senin için adımı kaybettim
Senin için arzuları kaybettim.
Bir tek ruhum önemli vardı bana,
Senin için ruhumu bile feda ettim.

Hayat ne kadar zor olursa olsun,
Şair ben, yaralarımla yaşıyorum.
Ama ben artık şair olmak değil,
Sevgili olmak istiyorum.

The reason you don't understand me, is not because you don't know the language - but because you don't know the person. When you start to know the person, language becomes secondary.

What this means is that - till your perception is empty of yourself, you can never understand another person for who they are. The same holds true for understanding the universe. Till you sever all ties with baseless assumption, you can never see the universe for what it is - instead, you'll always see it as you wish it to be. Wishfulness is an obstacle to understanding - wishfulness is an obstacle to knowledge.

Knowledge is love, and in love there is no wishing, only acting - there is no praying, only persevering - there is no silent spectating, only actively intervening.

Intervention is ascension, indifference is extinction. Prayerism is self-preservation, activism is world-preservation.

9. It's All You

It's All You
(Sonnet 1199)

There is no God but you,
There is no demon but you.
There is no future but you,
There is no past but you.

There is no time but you,
There is no space but you.
There is no distance but you,
There is no bridge but you.

There is no creation but you,
There is no creator but you.
There is no anchor but you,
There is no armageddon but you.

Mind is the construct, mind is constructor.
Either you are almighty, or you are a custard.

10. Experience is Crucial

How do you become almighty - by learning from every single situation of your life, and then applying that experience to expand your mind further, which in turn expands your life further.

Experience is crucial to understanding - you mustn't cook up your convictions solely based on second-hand experience and second-hand data. We still have to move forward stepping on previously published data, but make sure, you are never too blind to tell good data from bad data.

Advancement in science is predicated on clear thinking - clear thinking as in unbiased thinking. And neurologically speaking, although the mind can never be fully unbiased, with enough resolve, composure and conscience, you can think and behave less biased. And this very drive to behave less biased is what sets a real human apart from the crowd of human-looking animals.

But the problem is, it's quite simple to be human, but quite difficult to be simple. Simplicity is a breeding ground of answers and solutions. The reason that most lives are racked with questions and problems is because most deem simplicity to be unusual and clutter to be normal. Master simplicity, you'll master happiness. Master simplicity, you'll master life.

11. Simple Sonnet

Basit, Şiir

Barış doğar basitlikten,
Korku karmaşıklıktan.
Sükunet doğar sabırdan,
Belalar bencillikten.

Ama dertsiz hayat nasil hayat,
Zorluk bize cesaret getirir.
Karanlık hiç de ayıp değil,
Karanlık bize parlamayı öğretir.

İnsan korkmazsa kim korkacak -
Sebebi doğruysa korkmak ayıp değil.
Servetin derdi yalan derdidir,
Dünyada gerçek dertler eksik değil.

Dertte olduğunda herkes gözyaşı dökmeyi bilir,
Ama başkasının gözyaşını silmek için
derdine katlanmak, çok azı bilir.

Sonnet 1200

Simple, Sonnet

Serenity is born of simplicity,
Insecurity is born of clutter.
Patience empowers perseverance,
Selfishness brings down disaster.

But what's a life without difficulty,
Difficulty delivers durability.
Don't be ashamed of darkness in life,
It's in darkness we shine most brightly.

There's nothing shameful about fear,
It's a problem when the reason is baseless.
Trouble of privilege is trouble of lies,
Reject all privilege and rush to the helpless.

It is human nature to shed tears when in agony,
Taking pain to wipe another's tears is humanity.

12. No Harvest Without Rain

Acısız amacı yok,
Amaçsız ilacı yok.

Ain't no purpose without pain,
Ain't no harvest without rain.
Ain't no voice without silence,
Ain't no gallantry without gail.

Gözyaşlarından güneş doğar,
Yaralardan yazar doğar.

Tears are the gateway to sunrise,
Wounds are the gateway to insight.
Clarity comes by trodding on confusion,
Present unfolds at the end of yesternight.

Insightless sight is of no use,
Heartless head is of no use.
Voiceless lips are of no use,
Motionless limbs are of no use.

13. Himalayana

Himalayana
(Sonnet 1201)

Flashlights there are plenty,
Sun there is but one.
Rushmores there are plenty,
Everest there is but one.

Rushmore leers as a criminal monument,
While Everest stands as a gentle giant.
Arrogance goes together with brutality,
While valor and virtue go hand in hand.

Before you pursue intelligence,
Learn to behave like a human being.
Compared to an intellectual animal,
A lay person is the higher being.

The sun doesn't know what is radioactivity,
Yet it is our earth's radioactive messiah.
The Himalayas don't know fancy ideologies,
Yet ideologies crumble to grand Himalayana.

14. Benevolence Over Violence

Hatelessness is predicated on non-judgment. However, there is no such thing as perception without judgment. Then what does it mean, when one says, "I don't approve of hate"? It means that, I refuse to judge people based on prehistoric excuses of race, religion, gender and sexuality.

Just because the brain cannot live without judgment, doesn't mean it has to comply with every single judgment spit out by the primitive instinct of self-preservation.

Only cure for chronic cruelty is chronic kindness. Kindness can never overpower cruelty in the world unless we make it our first nature on purpose. And how do we make kindness our first nature? By simply not giving in to our actual evolutionary first nature of cruelty.

Evolution is a long and tedious process, but till now no species have had the brain capacity to consciously chart the course of its own evolution - until humankind came into the picture that is.

However, since we share the same origin of doom and destruction as the animals, till this moment acts of humanness don't come as seamlessly and naturally as animalness does.

Because deep down we still remain the same old wrath-hungry monsters of the nocturnal bushes.

This behavior is observed most clearly in the so-called modern world's social media behavior. Far more than facts and love, animal mind behind the human exterior is drawn to discrimination and disinformation.

The animal brain is neurologically wired to be more invested in violence and excitement than benevolence and boredom, because as animals our ancestors were exposed to violence and viciousness all their life, as opposed to the newly developed mental construct of kindness.

Hence the human mind's animal tendency of death and destruction manifests through the modern phenomenon of doomscrolling.

Doomscrolling is nothing new, people used to do the same with tv remote, switching channel after channel, rarely settling on any one program. And heads buried in social media news feed is nothing new either - before smartphone and internet heads used to be buried in actual physical newspapers. Only the means have changed, not the habit. This is not advancement, it's recurring derangement. I'll call it progress when you put down your phone or remote and actually listen to another person. Sure, phones can be a supplement to organic conversation, but never a replacement.

Apparently, the human mind is more inclined to explore secondhand descriptions of the world, than explore the world by themselves. When this changes, harmony, understanding and acceptance will be a natural phenomenon rather than one imposed by petty, synthetic geopolitical interventions.

ABHIJIT NASKAR

15. Green Energy

Till civic intervention becomes the norm, political intervention will achieve nothing. Till common sense becomes the normal sense, prejudice and superstition will keep festering.

However, the matter of civic intervention is not as black and white as it might seem - particularly because, in the absence of common sense society can't tell the difference between civic duty and mob violence. To put it simply, civic sense without common sense leads to mob violence.

Let's take climate activism for example.

Climate Activists (not all) have turned into Climate Karens, which has done nothing for the climate crisis, but has only added one more crisis to the list. BLM activists don't go about abusing white people, Pride activists don't go about harassing straight people, and yet, that's precisely what has become the norm in climate protests.

Vandalism isn't activism, you morons! If you want to help the climate, help the green energy industry become mainstream.

When it comes to consumer products, to the people of earth it's all about convenience, nothing else. Make green energy convenient, and fossil fuel consumption will sink to the bottom within weeks.

Think about Nokia - once upon a time Nokia was the king of the mobile communication industry. But with the launch of the iPhone with a convenient touch screen interface it didn't take long till Nokia was history. Apple wasn't the first to create the touchscreen smartphone though - IBM's Simon Personal Communicator, released in 1994, was the actual first handheld touchscreen smartphone. However, it was not until 2007 that the smartphone industry took off when Apple made the touchscreen interface clean, convenient and internet-friendly with their first iPhone.

Do something for the green energy industry like Apple did for the smartphone industry, and Nokia-like fossil fuel industry will fade away like passing clouds.

16. Commercialism Ain't Wrong

The best way to break a paradigm is not to fight it, but to make it obsolete. And how do we make a paradigm obsolete? By setting course for a new one - a civilized one.

So, don't fight the old, just stand as a beacon of the new - and the old will become irrelevant. Irrelevance kills a paradigm, not infamy. Relevance reinforces a paradigm, not popularity.

That's why I don't do paid promotions for my books - and so long as I am the sole owner of the rights, I never will.

Heck, I never even ask you to buy my books - I am not a writer, I am a revolution. If you wanna assimilate that revolution into your life, that oughta be your decision.

Everest doesn't pay climbers, Naskar doesn't pay readers. With every new publication all I say is "now out" - no promotions, no press, no media exposure, nothing - just "now available".

There are things that could be commercialized and should be commercialized, then there are things that must never be commercialized - for example, human rights.

Some might wonder, then why are my works locked behind a pay wall! To which I say, they are not - a vast portion of my work, including some of my most substantial sonnets and excerpts, are accessible freely over the internet. However, the only reason my entire body of work is not available for free is because bookstores do not approve of free books.

But, let me make one thing clear. There is nothing wrong with commercialism - problem is, in a world rooted in commercialism, human dignity, human principles, human virtue, human character get often trampled in the pathetic, single-minded pursuit of currency.

Thus faith gets corrupt, law gets corrupt - and yes, science too gets corrupt.

Hence, first and foremost we gotta pour some human dignity in our exploits of commercialism.

17. Virtue and Valor

I come from a working-class family, like most giants born and raised in the third world do – and this laborer's son had to make a name for himself by himself - only then his words earned the pedestal of honor where they are today.

The world doesn't care for the weak, it never has - so, first and foremost you gotta become strong - you gotta become a beacon, only then your words will have worth - only then your virtue will have value.

Be strong my friend, be valiant! It doesn't matter where you come from - what matters is what you are capable of.

Do science, do poetry, do theology - do whatever you like - but do it like a giant, not some puny, backboneless vermin.

Remember, virtue without valor is of no use - love without valor is of no use - science without valor is of no use.

Because, in the absence of valor, no matter how honorable your intentions are, society will trample all over you like doormat.

However, on the other hand, too much valor is equally dangerous. Unmoderated valor has a tendency of taking over the vision. Which means, valor is supposed to be wielded solely as a tool, never as the driver.

Valor should be servant to virtue, not the other way around. You know why? Because underneath the fancy word, valor is just brute force, which on its own does more damage than good. Brute force does't know right and wrong, that's why the compass of virtue is quintessential.

But here's the thing - virtue is much more than a compass - virtue is the bedrock of civilized life - better yet - virtue is the bedrock of civilization.

If that is indeed the case, we must ask one simple question.

What is virtue?

Is it merely the ability to tell right from wrong?

No.

Virtue is not the ability to tell right from wrong, virtue is the capacity to pursue the inconvenient right defying the comfort of the convenient wrong.

Which means, valor is intrinsic to virtue. But this is possible only with virtue born of realization, not those merely consumed theoretically. Theories have a place sure, but only as aid to realization, not the driver of realization.

Why?

Because only the faculties born of realization pass the test of time, all others fade just as quickly as they come about.

18. Beyond Fast and Easy

Don't run after fast fruits - brave the difficulty, persevere - only then the resulting life will have some meaning - only then the resulting realization will have substance.

More we persevere, more we realize - more we realize, more we exist. Fast and easy life facilitates a shallow existence, far worse than the animals. And yet, the entire world is going bonkers over such fast and easy exploits of the senses - fast and easy love, fast and easy food, fast and easy wealth.

There is more to life than sex, food and money - but alas, by the time they realize this they already are on their deathbed.

Change your ways, my friend - change while there is time. Later will be too late. Change your ways in love - change your ways in wealth - change your ways in life.

Move past the surface of things and take a look inside (no double talk intended). When you do, you shall realize - love is not about finding someone, love is about finding yourself in someone. Humanity isn't about finding humans, humanity is about finding yourself among the humans.

Life is about variation not hierarchy. In becoming inclusive we become human - in integration we become human.

All through history every culture on earth has produced its distinct literature - American literature, British literature, Latino Literature, Arabic literature, Turkish literature, European literature, Sanskrit literature, Bengali literature and so on. I am none of these, because I am all of these - Naskar is the amalgamation of all of world's cultures. Naskar is the first epitome of integrated Earth literature - where there is no inferior, no superior - no greater, no lesser. Soulfulness of Rumiland, heartfulness of Martíland, correctiveness of MLKland, sweetness of Tagoreland - merge them all in the fire of love, and lo emerges Naskarland - merge them all in the fire of love, and lo emerges lightland.

Now to the important bit.

In the making of light, brain is fundamental, while the heart is foundation. And nowhere this is more evident than in the exploits of science.

19. What is Illumination

Brain alone can't do science, if it could, you wouldn't need scientists, computers would suffice. You need a strong sense of humanity. And that's where the greatest of scientists stand out from the crowd of highly intellectual yet humanly mediocre scientists.

Let me put this into perspective.

You can be superior to me in intellect, but you can never be my superior in humanness – with effort you can be my equal, but never my superior - because I am the very height of humanness.

Here's the thing - even a little intellect when driven by great humanness achieves a greater impact on the world than a truckload of intellect with only a few sprinkles of humanness.

That's why I say, there is no glory in being the smartest person in the room - real glory is to be the rarest person to use their intellect for humanitarian purposes.

The world needs warmer scientists, not cleverer computers. The world does not need server after server full of meaningless data, what the world needs is the humanitarian drive to use the littlest of data for collective good.

What good is data, if it does no good!

What good is truth, if it does no good!

What good is knowledge, if it does no good!

Truth that does no good is worse than lie - knowledge that does no good is lower than ignorance. Illumination - illumination - illumination - that's what is needed.

But what is illumination?

Is it just discovery?

No.

Discovery is the simpler half of illumination, the braver half is application. Or to put it plainer still – discovery is the adolescent part of illumination, application is the adult part.

20. Humanitarian Science 101

Humanitarian Science 101
(Sonnet 1202)

BRAIN means Benevolent Reformer
Applying Information Nobly.
DATA means Determined Action
of Transformative Awareness.

Information Technology is primitive IT,
Civilized IT means Informed Transformation.
Heuristic and holistic can never go together,
Shortcuts only obstruct the rise of realization.

Electronics means electron artistry.
Chemistry is an art of bonding.
Mathematics is the art of numbers,
Evolution is the art of correcting.

Society without science dumps the world in stoneage,
Science devoid of society shoves the mind into iceage.

21. What Good is Data

Sonnet 1203

Head devoid of heart
shoves the mind into iceage.
Science driven by heart
is the strongest love language.

Science is humankind's greatest power,
Gravitas of which people are yet to fathom.
We let science be exploited by greed,
Rather than using it for universal reform.

Powerless heart and heartless power,
are both equally meaningless.
Dataless mind and mindless data,
are both equally dangerous.

What good is data if it does no good!
Science is born to serve the people,
not to lick the billionaire's boots.

22. What Good is Brain

Sonnet 1204

Data that does no good
is data of the dead.
Science that does no good
is science most savage.

Philosophy that does no good
is philosophy inane.
Theology that does no good
is theology in vain.

Faith that ends no division
is but faith of degradation.
Innovation that ends no suffering
is innovation towards armageddon.

What good is brain if it can't bring dawn!
What good is truth if it can't ease the mourn!

23. Kind and Clever

Kind and Clever
(Sonnet 1205)

A clever brain may not be kind,
A kind brain may not be clever.
When clever and kind come together,
That's the mark of a genius humanizer.

Plain science won't do no more,
What's needed is humanizing science.
Greater good can no more be collateral,
But the prime directive of all science.

Cold and clever ruins the world.
Cold and dumb is ruined with the world.
Kind and dumb changes with the world,
Kind and clever changes the world.

Study of science is easy,
Execution makes the difference.
Where there is no love for society,
there is no real use of science.

24. Harness Your Power

Sonnet 1206

Take your time to harness your power,
Then bring out that power for greater good.
Years in making are years of mocking,
Don't lose heart to the rage of the rude.

When I wrote my first book,
It took quite some effort and thinking.
But after about ten books or so,
Writing became natural as breathing.

Now whatever I speak is science,
Whatever I pen is poetry.
Whatever I think is philosophy,
Whatever I sense is psychology.

I never had an academic mentor,
I learnt everything by trial and error.
I was the first to recognize my own light,
Once I did, world gained a transformer.

25. Genius

Sonnet 1207

Transformers transform the world,
But who transforms the transformer?
Teachers rarely have the eye to find genius,
The genius must be their own illuminator.

My childhood showed no sign of genius,
I was pretty average in every task.
Yet today, Naskar is indelible
from the psyche of the universe.

I don't care for the word genius much,
I only care for the substance of human.
Only reason I mention the word genius,
is to rescue it from elitist exploitation.

Genius of head is nothing more than gimmick,
Genius of heart requires no such cheap trick.

26. Aim for Substance

Sonnet 1208

Genius of head stirs attraction,
Genius of heart stirs ascension.
Beauty of body stirs excitement,
Beauty of mind stirs illumination.

Put an end to all shallow pursuits,
For once in life, aim for substance.
Shallowness comes in many forms,
Sometimes appearance sometimes intelligence.

Intelligence is a good thing,
Only if it is guided by good.
Intelligence without a humane vision,
isn't worth being used even as sole of boot.

Genius of head may make some money,
Genius of heart makes humanity.
Be learned all you want in head,
Above all else develop some collectivity.

27. Mind Quotient

Mind Quotient
(Sonnet 1209)

Throw away all stupidity of IQ and EQ,
They are but stain upon mind's honor.
To quantify intelligence is stupid,
To quantify emotion is even stupider.

When the feeble psyche seeks reassurance,
It craves comfort in all sorts of nonsense.
Most times it resorts to the supernatural,
Exhausting that it resorts to pseudoscience.

It is no mark of mental progress to replace
supernatural bubble with pseudoscience bubble.
No matter how they try to sell you security,
Know that, human potential is unquantifiable.

IQ is no measure of intelligence,
EQ is no measure of emotion either.
But craving for IQ and EQ is symptom
of a shallow and feeble character.

28. The Journey

Sonnet 1210

Wealth can be measured,
Bodily beauty can be measured,
Online popularity can be measured,
And those who run after these measures
are but ghosts of our animal past.

Like BMI is an obsolete measure of health,
IQ is an outdated measure of intelligence.
With new evidence, discarding old constructs
is quintessential to the scientific process.

Science is a process of elimination,
Elimination of outdated methods and data.
Evolution is science, science is evolution,
Evolution through correction, not towards utopia.

Science ain't no journey from ignorant to learned.
Science is a journey of becoming less ignorant.

29. Timeless Scientist

Timeless Scientist
(Sonnet 1211)

Science is not a process of increasing knowledge,
Science is a process of decreasing ignorance.
But only a real scientist can fathom this,
While amateurs are intoxicated by evidence.

Evidence is crucial indeed, without question,
More crucial is the vision beyond evidence.
If you are too clingy to the truth of today,
You would make a lousy vessel for science.

Truth is a temporal spectrum,
Each slice contains a moment's evidence.
Bring the slices and string them together,
Lo, you develop the insight of science.

Mediocre scientist discovers a slice,
and feels all pleased with themselves.
Timeless scientist studies slice after slice,
and yet never stops feeling restless.

ABHIJIT NASKAR

30. Science Education

Science Education
(Sonnet 1212)

Science makes you restless,
Science destroys your sleep.
Science is ever inconvenient,
Science destroys your peace.

If you can persevere through all this,
Then you shall be a rightful scientist.
Otherwise you better go for a 9 to 5,
Science ain't for comfort seeking elitists.

Math, Bio, Physics, Chem and Code,
These five can make or break society.
Politics may be rooted in party loyalty,
Science is but ruined by field loyalty.

No field is an island entire of its own.
Those who can't collaborate,
for them the bell tolls.

31. Life Before Science

Sonnet 1213

If you are not offending some people,
You are not doing science.
Then again, if you put people in harm's way,
You are not doing science.

There is always some risk in science,
Often we take it for granted.
In the mindless pursuit of knowledge,
Scientists lose touch with humanness.

No science is worth more than human life,
No science is worth more than human dignity.
No science is worth more than human laughter,
No science is worth more than social safety.

History of science is full with horror stories.
No matter the past, modern science
must step with conscience and dignity.

32. We Are The Scientists

We Are The Scientists
(Sonnet 1214)

Justifying human rights violation as
necessary evil may be habit of politicians.
Scientists must be wiser than that, otherwise,
Science is just a weapon of mass destruction.

Scientist without humanity is anything but scientist,
Science without humanity is anything but science.
Civilized scientists work for the progress of humanity,
Primitive scientists work for the progress of science.

Progress of science is not necessarily progress of humanity,
Particularly when science advances trampling human life.
World leaders may brush off such matter as collateral,
To a scientist with spine nothing is higher than human life.

Whole world is in our care, beyond all law and politics.
We are capable, we are accountable - we are the scientists!

33. Scientists Ain't Politicians

If politicians had to live up to the same performance standards as scientists do, there would be no politicians in the world. We scientists don't have the luxury to make baseless claims and pompous promises, because we are too busy solving actual problems.

Of course there are exceptions in politics as well. But the point is, no other crucial field in the world thrives on the absolute disregard for expertise than politics. In fact, the more incompetent the better - that way they can be easily maneuvered by special interest groups. Sure, there are some good politicians, but that's all they are - good politicians. I am yet to see any radical initiative from any of them to curb the infestation of incompetence in the paradigm of politics - I am yet to see any real drive in any of them to reform the paradigm that entertains incompetence.

However, this particular work is meant particularly for scientists, so I have tried not to mention politics. The reason I am mentioning it here in the climax, is to point out the huge responsibility you carry on your shoulders as scientists. Unlike politicians, scientists actually gotta stand accountable - otherwise, the entire civilized fabric of society will fall apart in a matter of weeks.

If politicians disappear from the world, civil servants could easily do their job, because in reality, they already do. But if scientists disappear from the world, so will the world.

34. Science and Good

Sonnet 1215

Science is the strongest vanguard of life,
Science is the sanest voice of life.
Science is the calmest chord of life,
Science is the ablest keeper of life.

In ability there's nothing higher than science,
In worth there's nothing higher than life.
That makes science the greatest force for good,
Once it realizes its rightful duty to life.

Science is not god, humans are,
We gotta decide how we put it to use.
If science and good don't go together,
Science is but witchcraft of apes in suits.

Science begins where selfishness ends,
There's nothing lower than a selfish scientist.
Science takes us high when our head is held high,
With an unbent backbone driving good deeds.

35. Science Anthem

Science Anthem
(Sonnet 1216)

Science is my ode to society,
Science is serenade to society.
Science is the road to society,
Science is my aid to society.

Science is my poetry,
Science is philosophy.
Science is my thriller,
Science is my love story.

Science is not a love of knowledge,
Science is love of the light of knowledge.
Light of knowledge doesn't allow inhumanity,
Even if it's peddled in the benefit of knowledge.

Science is love, science is light,
Science is torch to the world at night.
Sometimes boring, sometimes daring,
Science es el loco amante of life.

36. 100 Questions of Life

1. What is life?

Life is the spirit to surpass life.

2. What is spirit?

Spirit is the drive that defines a human.

3. What is a human?

Human is the origin of civilization.

4. What is civilization?

Civilization is the faculty of correction.

5. What is correction?

Correction is perfection in motion.

6. What is perfection?

Perfection is imperfections we've made peace with.

7. What is peace?

Peace is the absence of apathy.

8. What is apathy?

Apathy is the end of senility.

9. What is senility?

Senility is accountability in practice.

10. What is accountability?

Accountability is fruition of awareness.

11. What is awareness?

Awareness is mental electricity.

12. What is the mind?

Mind is nature's gift of response.

13. What is nature?

Nature is order within chaos.

14. What is order?

Order is but friendship with chaos.

15. What is chaos?

Chaos is order we are yet to understand.

16. What is understanding?

Understanding is the will to expand.

17. What is expansion?

Expansion is triumph over rigidity.

18. What is rigidity?

Rigidity is paralysis.

19. What is paralysis?

Paralysis is ignorance in charge.

20. What is ignorance?

Ignorance is the key to knowledge.

21. What is knowledge?

Knowledge is ignorance we've chosen to correct.

22. What is choice?

Choice is the fulcrum of freedom.

23. What is freedom?

Freedom is the fulcrum of responsibility.

24. What is responsibility?

Responsibility is the act of backbone.

25. What is backbone?

Backbone is more than a stick to hang your head.

26. What is the head?

Head is the mightiest carrier of progress.

27. What is progress?

Progress is much more than mere functioning of nuts and bolts.

28. What are nuts and bolts?

Nuts and bolts are our greatest defense against unforeseen terrors of nature, on earth and beyond.

29. What is earth?

Earth is homeland of the humans.

30. What is a home?

Home is a people, not a place.

31. What are people?

People are potential immeasurable.

32. What is potential?

Potential is possibility despite instability.

33. What is possibility?

Possibility is the decision of humans.

34. What is a decision?

Decision is the will of unsubmission.

35. What is unsubmission?

Unsubmission is choosing honor over convenience.

36. What is honor?

Honor is conscientious narcissism.

37. What is narcissism?

Narcissism is destroyer of worlds when let loose, but builder of worlds when wielded by conscience.

38. What is conscience?

Conscience is synergy of heart and head.

39. What is the heart?

Heart in chest pumps blood, heart in brain pumps impulse.

40. What is the brain?

Brain is the seat of the being.

41. What is a being?

Being is protoplasm trying to escape protoplasm.

42. What is protoplasm?

Protoplasm is precursor to history.

43. What is history?

History is a book of lessons.

44. What is a lesson?

Every mistake is a lesson.

45. What is a mistake?

Mistakes are truth in making.

46. What is truth?

Truth is but grip over lies.

47. What is a lie?

Anything that harms people is a lie.

48. What is a harm?

Harm is the abuse of power.

49. What is power?

Power is energy used for good.

50. What is energy?

Energy is the absence of distance.

51. What is distance?

Distance is the root of indifference.

52. What is indifference?

Indifference is the reign of animal heritage.

53. What is heritage?

Heritage, in moderation, is an aid to growth, unmoderated, poison.

54. What is growth?

Growth is the refusal to bend.

55. What is it to bend?

To bend is to survive.

56. What is survival?

Survival is the reign of death.

57. What is death?

Death is but the fear of life.

58. What is fear?

Fear is but memory of our animal past.

59. What is memory?

Memory is the fabric of time.

60. What is time?

Time is the meaning behind moments.

61. What is meaning?

Meaning is sight driven by purpose.

62. What is sight?

Sight is a faculty that requires much more than eyes.

63. What are eyes?

Eyes are obstacle to real vision.

64. What is vision?

Vision is the drive to surpass the veins.

65. What are veins?

Veins are either vessel of valor or icewater.

66. What is valor?

Valor is rejuvenation through resilience.

67. What is resilience?

Resilience is tenacity of sweat.

68. What is tenacity?

Tenacity is failure tamed.

69. What is failure?

Failure is ambition in motion.

70. What is ambition?

Ambition is the herald of dawn despite the most evident dusk.

71. What is dawn?

Dawn is a symphony of intention and action.

72. What is intention?

Intention is the precursor to will.

73. What is will?

Will is but womb to the world.

74. What is the world?

Whole world's a mirror, we are but light.

75. What is light?

Light is but elixir born of wounds.

76. What is a wound?

Wounds are the gateway to wonder.

77. What is wonder?

Wonder is an exercise in curiosity.

78. What is curiosity?

Curiosity is a challenge to superstition.

79. What is superstition?

Superstition is nature's antidote to the insecurity of the unknown.

80. What is insecurity?

Insecurity is wisdom of the jungle against possible predatory attack.

81. What is wisdom?

Wisdom is the result of travel in mind, not in time or space.

82. What is travel?

Travel is the movement of mind, not limbs.

83. What is movement?

Movement is the act of ascension, not mere activity.

84. What is ascension?

Ascension is friendship with the horizon.

85. What is horizon?

Horizon is an heirloom of the past.

86. What is an heirloom?

Heirloom is a reminder of roots, not binder of chains.

87. What are roots?

Roots are a book of history, not of the future.

88. What is the future?

Future is an act of imagination.

89. What is imagination?

Imagination is war against the impossible.

90. What is impossibility?

Impossibility is invitation to a path yet unlaid.

91. What is a path?

Path is the defiance of defeat.

92. What is defeat?

Defeat is the feat of cowardice, not failure.

93. What is cowardice?

Cowardice is to accept the fiction of death over the fact of life.

94. What is fiction?

Fiction is the inability to fathom facts due to their vast intricacies.

95. What is a fact?

Fact is a state of matter.

96. What is matter?

Matter is an image made by mind.

97. What is an image?

Image is a reflection of our wishes.

98. What are wishes?

Wishes are inaction of the shoulderless without the guilt.

99. What are the shoulders?

Shoulders are the bedrock of civilized sentience.

100. What is sentience?

Sentience is the herald of life.

BIBLIOGRAPHY

ABHIJIT NASKAR

Archer M., (2000), Being Human: The Problem of Agency. Cambridge University Press.

Adolphs R (2003) Cognitive neuroscience of human social behaviour. Nature Rev Neurosci 4: 165–178.

Adolphs R, Tranel D, Damasio AR (2003) Dissociable neural systems for recognizing emotions. Brain Cogn 52: 61–69.

Andresen, Jensine, and Robert Forman, eds. Cognitive Models and Spiritual Maps. Bowling Green, Ohio: Imprint Academic, 2000.

Bernstein R.J., (1971), Praxis and Action: Contemporary Philosophies of Human Activity. Philadelphia: University of Pennsylvania Press.

Bernstein R.J., (1976), The Restructuring Social and Political Thought.

Bogen, J.E.(1995a), 'On the neurophysiology of consciousness: Part I. An overview', Consciousness and Cognition, 4.

Bogen, J.E. (1995b), 'On the neurophysiology of consciousness: Part II. Constraining the semantic problem', Consciousness and Cognition, 4.

Bremner, J. D., R. Soufer, et al. (2001). "Gender differences in cognitive and neural correlates of remembrance of emotional words." Psychopharmacol Bull 35 (3).

Brothers, L. (2002). The social brain: A project for integrating primate behavior and neurophysiology in a new domain. In J. T. Cacioppo et al. (Eds.), Foundations in neuroscience. Cambridge, MA: MIT Press.

Buss, D. D. (2003). Evolutionary Psychology: The New Science of Mind, 2nd ed. New York: Allyn & Bacon.

Buss, D. M. (1989). "Conflict between the sexes: Strategic interference and the evocation of anger and upset." J Pers Soc Psychol 56 (5).

Buss, D. M. (1995). "Psychological sex differences. Origins through sexual selection." Am Psychol 50 (3).

Buss, D. M., and D. P. Schmitt (1993). "Sexual strategies theory: An evolutionary perspective on human mating." Psychol Rev 100 (2).

Chomsky Noam, (2016) Who Rules the World?

Churchland, P.S. (1986), Neurophilosophy (Cambridge, MA: The MIT Press).

Churchland, P.S. & Ramachandran, V.S. (1993), 'Filling in: Why Dennett is wrong', in Dennett and His Critics:

Demystifying Mind, ed. B. Dahlbom (Oxford: Blackwell Scientific Press).

Churchland, P.S., Ramachandran, V.S. & Sejnowski, T.J. (1994), 'A critique of pure vision', in Large- scale Neuronal Theories of the Brain, ed. C. Koch & J.L. Davis (Cambridge, MA: The MIT Press).

Crick, F. (1994), The Astonishing Hypothesis: The Scientific Search for the Soul (New York: Simon and Schuster).

Crick, F. (1996), 'Visual perception: rivalry and consciousness', Nature, 379.

Crick, F. & Koch, C. (1992), 'The problem of consciousness', Scientific American, 267.

d'Aquili, Eugene. "Senses of Reality in Science and Religion." Zygon 17, no 4 (1982) d'Aquili, Eugene. "The Biopsychological Determinants of Religious Ritual Behavior." Zygon 10, no. 1 (1975)

d'Aquili, Eugene. "The Myth-Ritual Complex: A Biogenetic Structural Analysis." Zygon 18, no. 3 (1983)

d'Aquili, Eugene, and Andrew Newberg. The Mystical Mind: Probing the Biology of Religious Experience. Minneapolis: Fortress Press, 1999.

Damasio, A. (1994) Descartes' Error: Emotion, Reason and the Human Brain. New York, Putnams.

Damasio, A. (1999) The Feeling of What Happens: Body, Emotion and the Making of Consciousness. London, Heinemann.

Darwin, C. (1859) On the Origin of Species by Means of Natural Selection. London, Murray.

Darwin, C. (1871) The Descent of Man and Selection in Relation to Sex. London, John Murray.

Dawkins, R. (1976) The Selfish Gene. Oxford, Oxford University Press; a new edition, with additional material, was published in 1989.

Dewhurst, Kenneth, and A. W. Beard. "Sudden Religious Conversions in Temporal Lobe Epilepsy." British Journal of Psychiatry 117 (1970)

Dewhurst K, Beard AW. Sudden religious conversions in temporal lobe epilepsy. 1970 Epilepsy Behav 2003

Devinsky O, Lai G. Spirituality and religion in epilepsy. Epilepsy Behav 2008.

E. Horvitz, "One Hundred Year Study on Artificial Intelligence: Reflections and Framing," ed: Stanford University, 2014.

Eckhart Meister, Selected Writings

Farah, M.J. (1989), 'The neural basis of mental imagery', Trends in Neurosciences, 10.

Freud, S. "Selected papers on hysteria and other psychoneuroses" Journal of Nervous and Mental Disease 1909.

Freud, S. "The Origin and Development of Psychoanalysis", 1910

Freud, S. "Psychopathology of everyday life", 1914

Freud, S. "Beyond the Pleasure Principle", 1920

Frith, C.D. & Dolan, R.J. (1997), 'Abnormal beliefs: Delusions and memory', Paper presented at the May, 1997, Harvard Conference on Memory and Belief.

Gay, Volney, ed. Neuroscience and Religion. Plymouth, UK: Lexington Books, 2009.

Gazzaniga, M. S. (1985). The social brain. New York: Basic Books.

Gazzaniga, M.S. (1993), 'Brain mechanisms and conscious experience', Ciba Foundation Symposium, 174.

Geschwind N. "Behavioural changes in temporal lobe epilepsy". Psychol Med. 1979.

Gellhorn, E., Kiely, W.F. "Mystical states of consciousness: neurophysiological and clinical aspects." J Nerv Ment Dis. 1972;154:399-405.

Gilbert SL, Dobyns WB, Lahn BT (2005) Genetic links between brain development and brain evolution. Nat Rev Genet 6.

Gray JA. The Psychology of Fear and Stress. 2nd ed. New York, NY: Cambridge University Press; 1988.

Gloor, P. (1992), 'Amygdala and temporal lobe epilepsy', in The Amygdala: Neurobiological Aspects of Emotion, Memory and Mental

Dysfunction, ed J.P. Aggleton (New York: Wiley-Liss).

Gross CG, Rocha-Miranda CE, Bender DB (1972) Visual properties of neurons in the inferotemporal cortex of the macaque. J Neurophysiol 35: 96–111.

Guevara Che, The Motorcycle Diaries, 1992

Hardy, G. H. (1940). Ramanujan. Cambridge: Cambridge University Press.

Hall, Daniel, Keith Meador, and Harold Koenig. "Measuring Religiousness in Health Research: Review and Critique." Journal of Religion and Health 47, no. 2 (2008)

Harris, Sam, Jonas Kaplan, Ashley Curiel, Susan Bookheimer, Marco Iacoboni, and Mark Cohen. "The Neural Correlates of Religious and Nonreligious Belief." PLoS One 4, no. 10 (October 1, 2009)

Halgren, E. (1992), 'Emotional neurophysiology of the amygdala within the context of human cognition', in The Amygdala: Neurobiological Aspects of Emotion, Memory and Mental Dysfunction, ed J.P. Aggleton (New York: Wiley-Liss).

Halligan PW, Fink GR, Marshal JC, Vallar G. 2003. Spatial cognition: evidence from visual neglect. Trends Cogn Sci.

Handbook of Emotions, Edited by Michael Lewis, Jeannette M. Haviland-Jones, and Lisa Feldman Barrett, The Guilford Press; 3rd edition (2010).

Hameroff, S.R. and Penrose, R. (1996) Conscious events as orchestrated space-time selections. Journal of Consciousness Studies 3(1), 36-53; also reprinted in J. Shear (ed.) (1997) Explaining Consciousness-The Hard Problem. Cambridge, MA, MIT Press, 177-95.

Harding, D.E. (1961) On Having no Head: Zen and the Re-Discovery of the Obvious. London, Buddhist Society.

Hardy, A. (1979) The Spiritual Nature of Man: A Study of Contemporary Religious Experience. Oxford, Clarendon Press.

Harre, R. and Gillett, G. (1994) The Discursive Mind. Thousand Oaks, CA, Sage.

Haugeland, J. (ed.) (1997) Mind Design II: Philosophy, Psychology, Artificial Intelligence. Cambridge, MA, MIT Press.

Hauser, M.D. (2000) Wild Minds: What Animals Really Think. New York, Henry Holt and Co.; London, Penguin.

Hilgard, E.R. (1986) Divided Consciousness: Multiple Controls in Human Thought and Action. New York, Wiley.

Hilton, E.N., Lundberg, T.R. Transgender Women in the Female Category of Sport: Perspectives on Testosterone Suppression and Performance Advantage. Sports Med 51, 199–214 (2021).

Hitler, Adolf. Mein Kampf, 1925

Hodgson, R. (1891) A case of double consciousness. Proceedings of the Society for Psychical Research 7, 221-58.

Hofstadter, D.R. and Dennett, D.C. (eds) (1981) The Mind's I: Fantasies and Reflections on Self and Soul. London, Penguin.

Holland, J. (ed.) (2001) Ecstasy: The Complete Guide: A Comprehensive Look at the Risks and Benefits of MDMA. Rochester, VT, Park Street Press.

Holmes, D.S. (1987) The influence of meditation versus rest on physiological arousal. In M. West (ed.)

The Psychology of Meditation. Oxford, Clarendon Press, 81-103.

Holmstrom, David. 1992, Christian Science Monitor

Holloway RL (1996) Evolution of the human brain. In: Lock A, Peters CR (eds) Handbook of human symbolic evolution. Oxford University Press, Oxford

Jeannerod M (1988) The neural and behavioural organization of goal-directed movements. Clarendon Press, Oxford.

Johnson-Frey SH, Maloof FR, Newman-Norlund R, Farrer C, Inati S, Grafton ST (2003) Actions or hand-objects interactions? Human inferior frontal cortex and action observation. Neuron 39: 1053–1058.

Jackson, F. (1982) Epiphenomenal qualia. Philosophical Quarterly 32, 127-36.

James, W. (1890) The Principles of Psychology (2 volumes). London, Macmillan.

James, W. (1902) The Varieties of Religious Experience: A Study in Human Nature. New York and London, Longmans, Green and Co.

Jansen, K. (2001) Ketamine: Dreams and Realities. Sarasota, FL, Multidisciplinary Association for Psychedelic Studies.

Jay, M. (ed.) (1999) Artificial Paradises: A Drugs Reader. London, Penguin.

Jaynes, J. (1976) The Origin of Consciousness in the Breakdown of the Bicameral Mind. New York, Houghton Mifflin.

Kandel, E. R. In Search of Memory: The Emergence of a New Science of Mind, W. W. Norton & Company (2007).

Kandel E. R. Schwartz JH, Jessel TM. Principles of neural sciences. New York; McGraw Hill, 2000.

Kanwisher, N. (2001) Neural events and perceptual awareness. Cognition 79, 89-113; also reprinted inS. Dehaene (ed.) The Cognitive Neuroscience of Consciousness. Cambridge, MA, MIT Press, 89-113.

Kihlstrom, J.F. (1996) Perception without awareness of what is perceived, learning without awareness of what is learned. In M. Velmans (ed.) The Science of Consciousness. London, Routledge, 23-46.

Kosslyn, S.M. (1980) Image and Mind. Cambridge, MA, Harvard University Press.

Kosslyn, S.M. (1988) Aspects of a cognitive neuroscience of mental imagery. Science 240, 1621-6.

Kjaer, Troels, Camilla Bertelsen, Paola Piccini, David Brooks, Jorgen Alving, and Hans Lou. "Increased Dopamine Tone during Meditation- Induced Change of Consciousness." Cognitive Brain Research 13, no. 2 (April 2002)

Kölmel HW. 1985. Complex visual hallucinations in the hemianopic field. J Neurol Neurosurg Psychiatry.

Koenig, Harold. "Research on Religion, Spirituality, and Mental Health: A Review." Canadian Journal of Psychiatry 54, no. 5 (May 2009)

Koenig, Harold, ed. Handbook of Religion and Mental Health. San Diego, CA: Academic Press, 1998

Kraepelin E. Psychiatry: A Textbook for Students and Physicians. New York, NY: Science History Publications; 1990.

Lauglin, Charles, John McManus, and Eugene d'Aquili. Brain, Symbol, and Experience. 2nd ed. New York: Columbia University Press, 1992

Lakoff, G. and M. Johnson (1999). Philosophy in the flesh. Basic Books: New York.

LeDoux, J. E. (1996). The emotional brain. New York: Simon & Schuster.

LeDoux, J.E. (1992), 'Emotion and the amygdala', in The Amygdala: Neurobiological Aspects of Emo- tion, Memory and Mental Dysfunction, ed J.P. Aggleton (New York: Wiley-Liss).

Levin, D.T. and Simons, D.J. (1997) Failure to detect changes to attended objects in motion pictures. Psychonomic Bulletin and Review 4, 501-6.

Levine,J. (1983) Materialism and qualia: the explanatory gap. Pacific Philosophical Quarterly 64, 354-61.

Levine,J. (2001) Purple Haze: The Puzzle of Consciousness. New York, Oxford University Press. Levine, S. (1979) A Gradual Awakening. New York, Doubleday.

Levinson, B.W. (1965) States of awareness during general anaesthesia. British Journal of Anaesthesia 37, 544-6.

Lewicki, P., Czyzewska, M. and Hoffman, H. (1987) Unconscious acquisition of complex procedural knowledge. Journal of Experimental Psychology: Learning, Memory and Cognition 13, 523-30.

Naskar, Abhijit. "What is Mind?", 2016

Naskar, Abhijit. "Love, God & Neurons: Memoir of A Scientist who found himself by getting lost", 2016

Naskar, Abhijit. "Principia Humanitas", 2017

Naskar, Abhijit. "We Are All Black: A Treatise on Racism", 2017

Naskar, Abhijit. "Either Civilized or Phobic: A Treatise on Homosexuality", 2017

Naskar, Abhijit. "Build Bridges not Walls: In the name of Americana", 2018

Naskar, Abhijit. "Citizens of Peace: Beyond the Savagery of Sovereignty", 2019

Naskar, Abhijit. "The Constitution of The United Peoples of Earth", 2019

Naskar, Abhijit. "Mission Reality", 2019

Naskar, Abhijit. "Good Scientist: When Science and Service Combine", 2020

Newberg, Andrew, and Jeremy Iversen. "The Neural Basis of the Complex Mental Task of Meditation: Neurotransmitter and Neurochemical Considerations." Medical Hypotheses 61, no. 2 (2003).

Newberg, Andrew. "How God Changes Your Brain: An Introduction to Jewish Neurotheology", CCAR

Journal: The Reform Jewish Quarterly, Winter 2016.

Newberg, Andrew, and Stephanie Newberg. "A Neuropsychological Perspective on Spiritual Development." In Handbook of Spiritual Development in Childhood and Adolescence, edited by Eugene Roehlkepartain, Pamela King, Linda Wagener, and Peter Benson. London: Sage Publications, Inc., 2005

Newberg, Andrew. "The Neurotheology Link An Intersection Between Spirituality and Health", Alternative and Complimentary Therapies, Vol 21 No 1, February 2015.

Newberg, Andrew, Nancy Wintering, Dharma Khalsa, Hannah Roggenkamp, and Mark Waldman. "Meditation Effects on Cognitive Function and Cerebral Blood Flow in Subjects with Memory Loss: A Preliminary Study." Journal of Alzheimer's Disease 20, no. 2 (2010)

Nash, M. (1995), 'Glimpses of the mind', Time.

Nesse RM. Proximate and evolutionary studies of anxiety, stress and depression: synergy at the interface. Neurosci Biobehav Rev. 1999;23:895-903.

Nicolelis, Miguel. (2011) "Beyond Boundaries: The New Neuroscience of Connecting Brains with Machines---and How It Will Change Our Lives", Times Books

O'Hara, K. and Scutt, T. (1996) There is no hard problem of consciousness. Journal of Consciousness Studies 3(4), 290-302, reprinted in J. Shear (ed.) (1997) Explaining Consciousness. Cambridge, MA, MIT Press, 69-82.

O'Regan, J.K. and Noe, A. (2001) A sensorimotor account of vision and visual consciousness. Behavioral and Brain Sciences 24(5), 883-917.

Ornstein, R.E. (1977) The Psychology of Consciousness (2nd edn). New York, Harcourt.

Ornstein, R.E. (1986) The Psychology of Consciousness (3rd edn). New York, Pehguin.

Ornstein, R.E. (1992) The Evolution of Consciousness. New York, Touchstone.

Penfield W, Faulk ME (1955) The insula: further observations on its function. Brain 78: 445– 470.

Penrose, R. (1994), Shadows of the Mind (Oxford: Oxford University Press).

Penrose, R. (1989), The Emperor's New Mind: Concerning Computers, Minds and The Laws of Physics (Oxford: Oxford University Press).

Persinger, "'I would kill in God's name' role of sex, weekly church attendance, report of a religious experience and limbic lability" Perceptual and Motor Skills 1997.

Persinger "Experimental simulation of the God experience" Neurotheology 2003.

Persinger, Corradini, Clement, Keaney, et al "Neurotheology and its convergence with neuroquantology" NeuroQuantology 2010.

Persinger. "The neuropsychiatry of paranormal experiences". J Neuropsychiatry Clin Neurosci 2001.

Persinger. "Neuropsychological bases of god beliefs", New York: Praeger, 1987

Persinger. "Temporal lobe epileptic signs and correlative behaviors displayed by normal populations", Journal of General Psychology, 1986

Perry BD, Pollard R. Homeostasis, stress, trauma, and adaptation. A neurodevelopmental view of childhood trauma. Child Adolesc Psychiatr Clin N Am. 1998;7:33.

Ramachandran VS. Behavioral and magnetoencephalographic correlates of plasticity in the adult human brain. Proc Natl Acad Sci USA 1993; 90: 10413–20.

Ramachandran VS. Plasticity and functional recovery in neurology. Clin Med 2005; 5: 368–73.

Rock I, Victor J. Vision and touch: an experimentally created conflict between the two senses. Science 1964; 143: 594–6.

Roberts, TA; Smalley, J; Ahrendt, D (December 2020). "Effect of gender affirming hormones on athletic performance in transwomen and transmen: implications for sporting organisations and legislators". British Journal of Sports Medicine. 55 (11): 577–583

Royet JP, Plailly J, Delon-Martin C, Kareken DA, Segebarth C (2003) fMRI of emotional responses to odors: influence of hedonic valence and judgment, handedness, and gender. Neuroimage 20: 713–728.

Rozin R Haidt J and McCauley CR (2000) Disgust. In: Lewis M, Haviland-Jones JM (eds) Handbook of Emotion. 2nd Edition. Guilford Press, New York, pp 637–653.

Saxe R, Carey S, Kanwisher N (2004) Understanding other minds: linking developmental psychology and functional neuroimaging. Annu Rev Psychol 55: 87–124.

S. J. Russell and P. Norvig, Artificial intelligence: a modern approach (3rd edition): Prentice Hall, 2009.

Singer T, Seymour B, O'Doherty J, Kaube H, Dolan RJ, Frith CD (2004) Empathy for pain involves the affective but not the sensory components of pain. Science 303: 1157–1162.

Smith A (1759) The theory of moral sentiments (ed. 1976). Clarendon Press, Oxford.

Schilling, Vincent. 2017, indian country today

Stein, Stephen K. 2017, The Sea in World History: Exploration, Travel, and Trade

Tesla N. "My Inventions", 1919

T. R. Society, "Machine learning: the power and promise of computers that learn by example," ed. The Royal Society, 2017.

Tomasello M, Call J (1997) Primate cognition. Oxford University Press, Oxford

·

Printed in Great Britain
by Amazon